RAINTREE
SCIENCE
ADVENTURES

CAVES

Judith E. Greenberg and Helen H. Carey
Illustrated by Yoshi Miyake

RSVP
RAINTREE
STECK-VAUGHN
PUBLISHERS
The Steck-Vaughn Company

Austin, Texas

For Mitch — **J.G.**
For Ryan — **H.C.**

Editorial
Barbara J. Behm, Project Editor
Judith Smart, Editor-in-Chief

Art/Production
Suzanne Beck, Art Director
Kathleen A. Hartnett, Designer
Andrew Rupniewski, Production Manager
Eileen Rickey, Typesetter

Reviewed for accuracy by:
Gretchen M. Alexander, Executive Director
West 40 Educational Service Center Number 5
Northlake, Illinois

Scott R. Welty, Instructor
Maine East High School
Park Ridge, Illinois

Copyright © 1990 Raintree Publishers Limited Partnership

Library of Congress Number: 89-78289

    4   5   6   7   8   9   94 93

Library of Congress Cataloging-in-Publication Data

Greenberg, Judith E.
  Caves / by Judith E. Greenberg and Helen H. Carey;
illustrated by Yoshi Miyake.
  (Raintree science adventures)
  Summary: Describes how different kinds of caves and cave formations are created in nature and what caves and cave animals are like. 1. Caves—Juvenile literature. [1. Caves.] I. Carey, Helen H. II. Miyake, Yoshi, ill. III. Title. IV. Series.

GB601.2.C37 1990 551.4'47   89-78289
ISBN 0-8172-3750-X (lib. bdg.)

Before You Begin

This book takes *you* on an adventure inside a dark and mysterious cave! Throughout your exploration, you will collect scientific information about the underground world of caves.

You will take part in an experiment that will help you see how caves are made. You will learn about cave formations and the creatures that live there in the dark.

Knowing the words below will help you in your adventure.

acid   a chemical that can dissolve metals and some types of rock. Acids have a sour taste. Some examples are lemon juice and vinegar.

biologist   a scientist who studies living things

canteen   a small container for carrying liquids

cavern   a large cave

cavity   a hole

compass   an instrument that shows direction

erosion   a wearing away by water, wind, or ice

geologist   a scientist who studies the earth, especially its rocks

glacier   a large mass of ice that moves slowly along the land

limestone   a kind of rock made up mostly of calcium carbonate

Now turn the page, and begin your science adventure!

You Find a Cave!

You are going on a camping trip with your best friends Hillary and Harry Nichols. They are twins. Their parents and older brother Jack are going, too. Everyone will have fun camping. You pack carefully because everything you will need has to fit in your backpack.

The group leaves early in the morning. You drive to Icicle Lake and arrive at the campground in time for lunch. You eat your sandwich quickly because you are excited about going exploring with Hillary, Harry, and Jack.

As you leave, their mother says, "Don't go too far. Listen to Jack, and come back in an hour."

You carry a backpack containing a flashlight, some rope, and a **canteen** of water. Hillary has a notebook and a pencil to draw any animals or flowers that she sees. Harry has a **compass** and a whistle.

The group's first stop is at the edge of the lake. Hillary finds a water bug to draw. Harry finds a smooth and pretty rock. He puts it in his pocket. Jack studies some animal tracks. You start walking up a nearby hill.

"Come see what I've found on this big rock," you say. You point to something that looks like spots of paint.

"I know what that is," says Jack. "It's called lichen. Lichens are made up of two types of plants—algae and fungi. Lichens grow on rocks and in caves."

You study the lichens for a while. Then suddenly you feel a cool breeze on your back.

The breeze is coming out of an opening in the side of the hill. "It's a cave!" you shout. "I found a cave!"

The others also look at the opening. You think about going in. The hole is big enough for you to crawl through. When you poke your head into the hole, it is dark and damp. You can hear water dripping from somewhere far inside the cave.

Harry shines the flashlight into the opening. Mud and rocks cover the cave floor. Even with the flashlight, you cannot see very far into the cave.

"Do you think it's a pirate cave? Maybe we can find a treasure chest!" Harry says excitedly.

"Maybe someone once got stuck in the cave and died. Maybe we'll find the skeleton!" Hillary says.

You think that exploring a cave would be fun. You wonder, however: What if you get lost in the cave? What if you get stuck? What if you fall in a deep hole? What if there are bears in the cave?

Jack says, "I've been in caves before. They're interesting to explore. I'll be your guide."

You get the rope out and tie one end of it to a big rock near the cave opening. You put the other end through the belt loops on each person's jeans. Harry gives the whistle to Jack.

Holding the flashlight, Jack leads the way into the cave. You all agree that if you hear the whistle blow two times, everyone should get out of the cave fast!

Once inside the cave, you shiver from the coolness. Jack explains that it's always cool inside caves because there is no sunlight to create heat. This also explains why caves are completely dark. You can only see where the flashlight shines. All of you crawl farther into the cave. Then you reach a place where the cave is big enough for you to stand up.

You can hear water dripping from the roof of the cave. You follow the sound.

Suddenly Hillary screams, "Something is flying around in here! There could be bats!" Jack blows the whistle twice. Everyone turns and scrambles out of the cave.

You arrive back at the camp excited about your cave visit. All of you try to talk at once.

"Whoa, slow down. Catch your breath," says Mr. Nichols. "Now, what's this about a pirate cave?"

Hillary explains where you found the cave. Harry says he thinks pirates have buried treasure inside the cave. You tell about the darkness, the coolness, the water, and the bats that could be inside the cave.

Mr. Nichols says, "You were smart to take Jack into the cave with you rather than to go by yourselves. Jack has been in many caves. He's a spelunker. Spelunkers are people who go into caves to study them. They have to learn how to explore caves safely. They need to have special equipment. They must learn and practice safety rules."

Mr. Nichols tells you that spelunkers wear warm clothes that are waterproof. They wear hard hats to protect their heads from falling rocks. Each day, a spelunker takes along a canteen of water and enough food for two meals. Exploring caves is very hard work, and spelunkers can get very hungry and thirsty.

"While Jack goes fishing tomorrow, the rest of you are going on a special tour of Icicle Caverns. Icicle Caverns is an underground **limestone** cave. A guide there will take you through the cave in a raft. You'll see just how fascinating the inside of a cave can be," Mr. Nichols says.

That night, you are so excited about exploring Icicle Caverns that you have trouble falling asleep.

Cave Secrets

The next day, Roy Ruiz, your guide, meets you at the entrance to Icicle Caverns. He is wearing gray coveralls and sturdy boots. He is carrying a backpack and a rope. A special sling goes across his chest and over one shoulder. The sling can be attached to a rope if he has to be lowered into a cave. His hard hat has a light on the front of it.

"Hi, kids," Mr. Ruiz says as he shakes your hands. "I'm glad you wore old clothes. Caves can be pretty muddy. I'm wearing all this caving gear so you can see what I look like when I'm exploring a new cave. These **caverns** have been explored and made safe for tourists. However, we like our visitors to wear special hard hats just to be extra careful."

"What kind of light is on our hats?" Hillary asks Mr. Ruiz.

"It's a carbide lamp, the kind that miners wear when they work underground in a mine. It's cheap and easy to use, and it weighs very little. The chemical in it is calcium carbide. It makes a gas when water from the top part of the lamp drips on it. When the gas comes through a little tube in the lamp, it mixes with air. The gas can then be lit with a match. The flame makes a bright light."

You and the twins can hardly wait to get into the cave. You ask Mr. Ruiz, "Where are we going first?"

"You're going to an underground lake!" he says. He laughs when he sees your look of surprise.

"We're going into the cave in a raft. Watch your step getting in," says Mr. Ruiz.

You sit down in the front of the small rubber raft. You reach over the side of the raft to feel the water. It is cold!

Mr. Ruiz's oar pushes the raft through the opening of the cave. You travel through a narrow, twisting passage. Sometimes the raft rubs against the walls of the cave as it moves. In some places, the walls are wet and cool like a melting icicle. In other places, the cave walls have sharp, pointy rocks.

You are glad that there are electric lights inside the cave. Without these lights and the lamp on your hard hat, you couldn't see your hand in front of your face. The cave air is cool and damp. It smells like wet mud.

Harry wants to know how this cave was made. Mr. Ruiz tells him that the rocks around you are as old as the world itself. However, he says the cave itself is probably only a few hundred years old. "If you listen very carefully," says Mr. Ruiz, "you'll hear something that will tell you how the cave was formed." Mr. Ruiz stops the raft, and everyone is quiet.

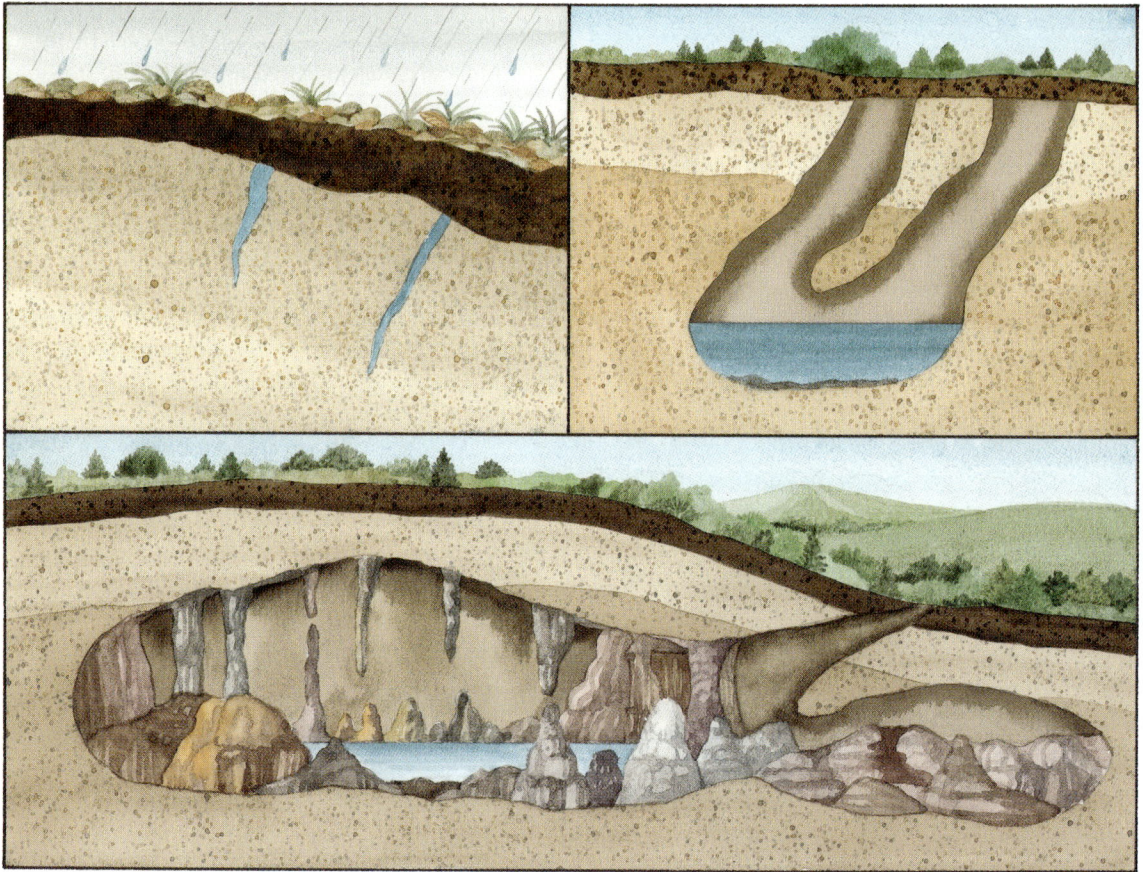

The only sound is water dripping from the rocks. "How does dripping water make a cave?" you ask.

Mr. Ruiz answers, "Water slowly wears rocks away. This is called **erosion.** The main shaping of the cave, however, was caused by water containing carbon dioxide. Carbon dioxide is a gas that comes from rotting leaves and the soil. The gas-filled water seeps down through cracks in the soft limestone rocks. The water slowly dissolves the calcium carbonate in the rocks, forming a **cavity.** After a long time, this cavity becomes a cave."

The raft travels out of the narrow passage into a large open area. It's the underground lake!

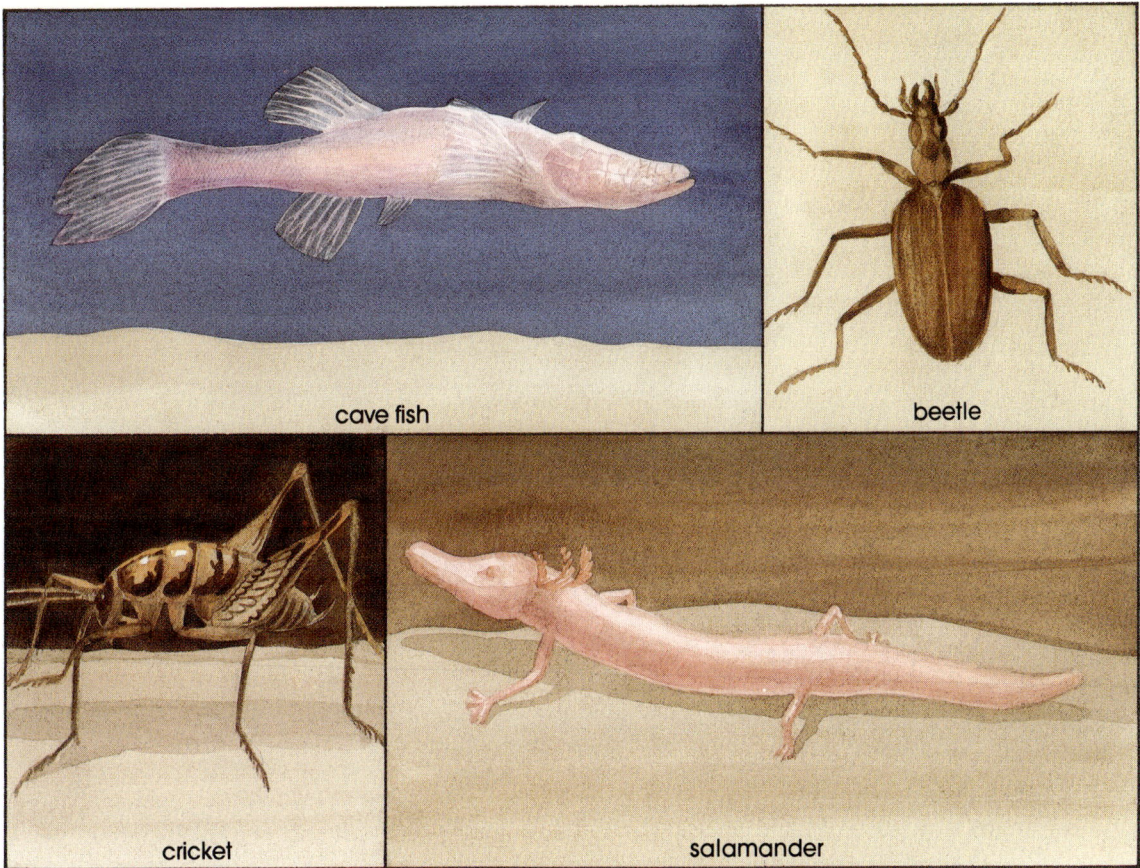

cave fish

beetle

cricket

salamander

Suddenly you feel your stomach flop. "There's something alive in the water!" you shout.

"Don't be afraid. This little thing won't hurt you. He can't even see you," Mr. Ruiz explains. He scoops up a finger-size fish in a net. "This blind cave fish and certain beetles, crickets, and salamanders live their entire lives inside caves. They are called troglobites, or cave dwellers."

You see that the fish has pale coloring and no eyes. Mr. Ruiz tells you that troglobites have excellent senses of smell, touch, and hearing. This helps them find food in the dark cave.

Mr. Ruiz also tells you that water does more than shape a cave. It also makes it beautiful. Mr. Ruiz brings the raft to the edge of the lake. Everyone gets out of the raft to take a closer look around. Mr. Ruiz says that you are looking at stalactites. They grow down from the ceiling of a cave. Stalactites are formed by water dripping from the cave roof. The water contains carbon dioxide, which dissolves the calcium carbonate found in the rock ceiling. Where water drips, it loses carbon dioxide into the air and leaves behind tiny crystals of calcium carbonate. The crystals build up very slowly to form stalactites. The process is so slow that it may take a hundred years for a stalactite to grow one inch.

Hillary is looking at stalagmites. Mr. Ruiz says that they grow up from the floor of the cave. Drops of water that land on the cave floor form them. The stalagmites grow upward as calcium carbonate crystals build up. Stalagmites are thicker at the bottom than they are at the top. Sometimes a stalagmite will join a stalactite to form a column.

Harry is looking at cave formations called "soda straws" and "curtains." A soda straw is a thin, hollow tube of calcium carbonate crystals. Water drips through this tiny tube. Cave curtains are formed by water flowing down a sloping ceiling or ledge. The curtains, made of calcium carbonate crystals, are about as thick as a person's thumb. You can see light through them.

Cave Experiment

Mr. Ruiz docks the raft and leads you to a place where visitors can buy bag lunches in the caverns. Exploring the cave has made you very hungry. You sit down with Harry and Hillary to eat. There is a sign on a flat rock near you. The sign reads, "CAVE EXPERIMENT."

You put your sandwich down and go to look. There are three cups on the rock. You ask Mr. Ruiz to tell you about the cups.

Mr. Ruiz says, "The guides have set up this experiment in the cave because it helps show how the cave was formed."

He puts eggshell pieces in each cup. He tells you that eggshells are made of calcium carbonate. This is the same substance that makes up limestone rocks.

He asks you to pour some plain water into cup number one.

He asks you to pour some club soda into cup number two.

He asks you to pour some vinegar into cup number three.

Tell what you think will happen to the eggshell pieces in each cup. Make a box like this on a piece of paper. Write your guesses in it.

1) eggshell in water	2) eggshell in club soda	3) eggshell in vinegar

"We will have to wait a day or so to see what happens to the eggshells," Mr. Ruiz explains. "I can show you the eggshells from two days ago, though."

You look at the eggshells that were in plain water and in club soda. The eggshells did not change. They look and feel the same as before. The eggshells that were in vinegar, however, are all soft and mushy. You rub them between your fingers. The eggshells leave white streaks on your fingers.

"Vinegar is a strong **acid**," Mr. Ruiz says. "It dissolved the calcium in the eggshell. The club soda is a mild acid. It will take more time before a change can be seen in that eggshell.

"The experiment shows what the acid carbon dioxide does to the calcium carbonate in limestone."

Different Types of Caves

"Do water and limestone make every kind of cave?" you wonder aloud.

Mr. Ruiz answers, "No. Some caves are made by volcanoes or by the wind. Sometimes burning lava comes out of a volcano. When the top layer of lava cools, it becomes very hard. Under it, hot lava is still flowing. When all the hot lava flows away, a cave is left under the hard top layer.

"Warm winds can hollow out a cave in a mountain of ice. These are called **glacier** caves. These caves are dangerous. Melting ice can make the cave fall in," warns Mr. Ruiz.

"Once we took a boat through a cave that the ocean waves made in the rocks," says Hillary.

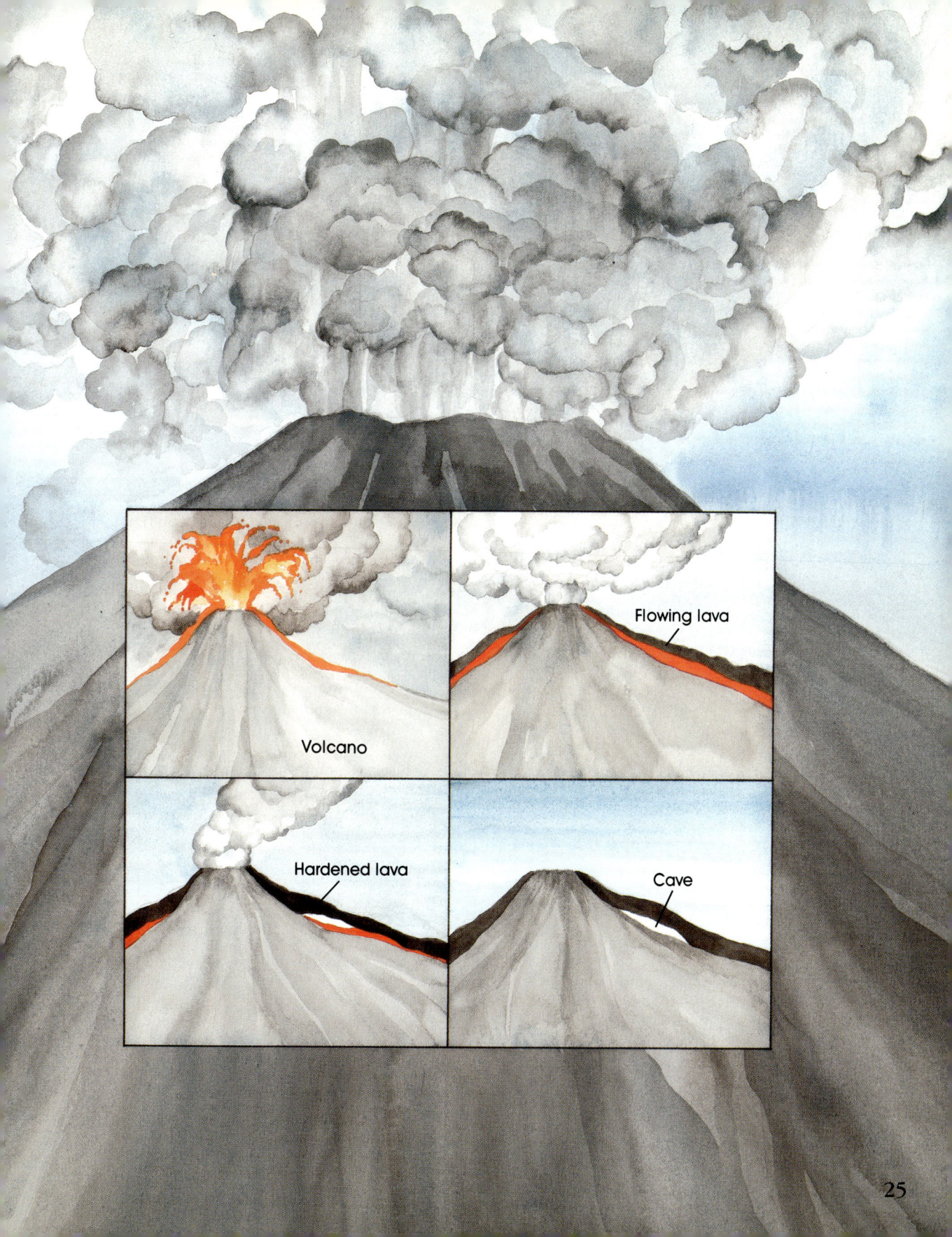

Volcano

Flowing lava

Hardened lava

Cave

"What kind of caves did cave people live in?" Harry asks.

"Good question," says Mr. Ruiz. "Throughout the world, ancient cave people and cave bears lived in caves made by the wind. Blowing wind carved out a hole in a cliff. These caves were good for shelter.

"In some cases, the ancient people drew pictures on the cave walls. These pictures show the animals that cave people hunted. Skeletons of the cave bears show that they were much larger than our bears today."

"Ooh, I hope we don't find any bear skeletons in this cave," you say, looking over your shoulder.

Mr. Ruiz explains, "I don't think there are any bears in here, but there are some animals that we could visit. Can you guess what they are? I'll give you a little hint. There are probably thousands of them."

The Bat Cave

You crawl, squeeze, and wriggle your way through a small passageway. Suddenly, the space is wider, and you are able to stand up. Mr. Ruiz shines his flashlight on the ceiling. Thousands of bats are hanging upside down from the cave roof.

"Bats have lived in this part of the cave for many years. They eat mostly insects and fruit. They hunt for these at night outside the cave," says Mr. Ruiz.

"How long can the bats hang like that?" you ask.

"In the winter, these bats hibernate, or become inactive. They hang from the ceiling until spring. During hibernation, the bats' temperature, heartbeat, breathing, and blood circulation all slow down. They don't eat while they are hibernating," says Mr. Ruiz.

Mr. Ruiz explains that cave bats look strange, but they will not hurt you. They can't see very well in the light outside the cave, so they sleep in the daytime. They fly at night without bumping into anything.

"How do they do that?" you ask.

Mr. Ruiz says, "Bats use echoes to discover where things are. A bat makes a sound with its mouth or nose that people are unable to hear. The sound bounces off whatever is in the bat's path. The sound then comes back to the bat's big ears. This echo lets the bat know the shape and location of objects. Therefore, the bats know what to stay away from and where there is food."

Indians in Caves

You follow Mr. Ruiz to another part of the cave. You see a real American Indian chief dressed in a deerskin shirt and leggings. He is wearing moccasins on his feet.

"I want you to meet Chief Bearclaw. He is the guide in this special part of Icicle Caverns," says Mr. Ruiz.

Chief Bearclaw shows you some very old paintings of animals on the cave wall.

You did a report on American Indians for school, but you didn't know that any of them lived in caves. "Why did Indians live here?" you ask the chief.

"Indians used this cave for shelter. This is how the cave looked when the Indians lived in it," says Chief Bearclaw. He points out the pottery, bows and arrows, and spears lying about.

The three of you walk around and look closely at the things the Indians used. You wish you had a bow and arrow like the ones in the cave.

Outside the cave, you thank Mr. Ruiz for leading you on the cave adventure.

"I had fun," says Hillary. "When I grow up, I'm going to be a **biologist** and study cave animals—even bats!"

"Not me," says Harry. "I'm going to be a **geologist** and study rocks."

What kind of science adventure would you like to go on next?